SCIENCE MUSEUM

The Steam Engine

A brief history of the reciprocating engine

By

R. J. LAW, B.Sc., A.C.G.I.

LONDON

HER MAJESTY'S STATIONERY OFFICE

1965

James Watt (1736–1819). From the painting by Sir William Beechey.

Introduction

With the disappearance of steam locomotives from railways at home and abroad, it is sometimes thought that the day of steam power is past. This is far from the truth. In those countries, such as our own, where thermal energy forms the basis of electric power supplies, steam is still the medium used for converting heat into mechanical energy to drive the electric generators. In industrialised countries electric power alone represents the equivalent of several slaves for every man, woman and child of the population. It is therefore not easy to appreciate the sheer drudgery that has persisted until quite recent times and still persists in undeveloped areas of the world. When the only prime movers were windmills and water wheels, most physical work was done by men supplemented by animals.

The steam engine was the agent which began the liberation of mankind from toil. It also made power available on a scale unknown before, leading to an unprecedented expansion of industry. The importance of its history therefore transcends its purely technical aspects, interesting though these are.

By the seventeenth century mines had become sufficiently deep for pumping to be a serious problem. Where water power could not be used, the pumps were worked by horses or men, but at great cost. In the same century the discovery of the pressure of the atmosphere suggested a new source of power, if only a vacuum could be obtained. At the beginning of the 18th century the first successful cylinder and piston steam engine was made. Its use became general in British coalfields, but throughout the century it remained entirely dependent upon the vacuum formed by the condensation of steam. In the nineteenth century the introduction of high pressure steam made mechanical transport possible on land as well as at sea. As the century progressed, steam became pre-eminent and supplied the power for every conceivable purpose from cotton mills to lawn mowers.

In the 20th century the reciprocating steam engine has been almost eclipsed by the internal combustion engine and by the steam turbine, which can utilise much higher steam pressures and develop far larger powers. However for small powers where the exhaust steam is required for heating or process work, the reciprocating steam engine is still to be found.

The Pressure of the Atmosphere

In the first century A.D. Hero of Alexandria described a device to open the doors of a temple by heat from an altar fire. The expansion of air, when heated, drove water out of a closed tank into a bucket, which descended and opened the doors. When the air cooled, it contracted and sucked the water out of the bucket, closing the doors.

In 1606 Giovanni Battista della Porta of Naples described a laboratory experiment in which steam instead of air forced water out of a closed tank. He also described how a flask full of steam with its neck thrust beneath cold water, would draw water up as the steam condensed. The accepted explanation of suction was that 'nature abhorred a vacuum'. In 1641 the engineers of Cosimo de' Medici, Grand Duke of Tuscany, attempted to make a suction pump to draw water from a depth of 50 feet. When it failed to draw from more than the usual depth of 28 feet, they appealed to Galileo for an explanation. He could only reply that there was evidently a limit to nature's abhorrence, but his interest was aroused. After his death the following year experiments were begun by his pupil Evangelista Torricelli, In 1643 Torricelli announced that the atmosphere exerted a pressure on account of its weight and that this pressure would balance a column of mercury about 30 inches in height. This discovery caused widespread interest and probably prompted Otto von Guericke, burgomaster of Magdeburg, to make the first air pump. About 1650 he succeeded in evacuating a copper sphere, which collapsed under the atmospheric pressure. In 1654 he demonstrated this pressure with a cylinder about 15 inches in diameter having an accurately fitted piston packed with hemp. By pumping the air out of the cylinder, he was able to raise a weight of more than one ton.

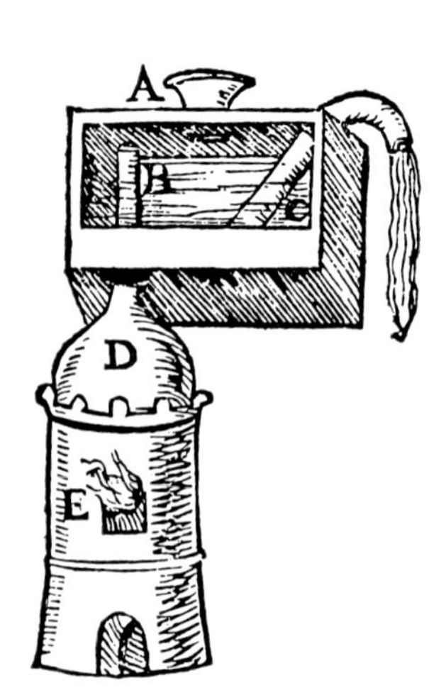

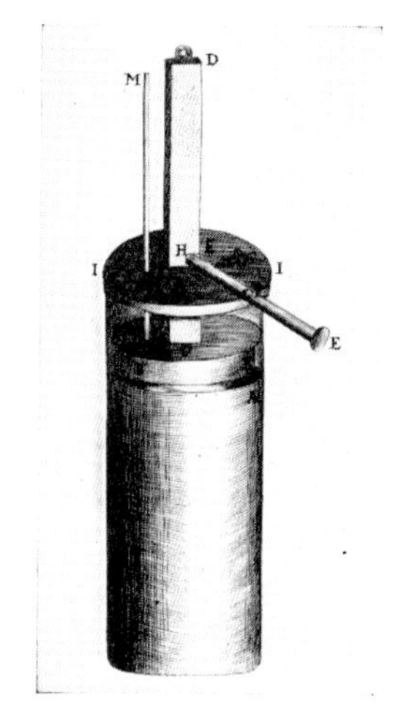

Above: Della Porta's steam pressure apparatus 1606.
Right: Papin's cylinder and piston apparatus 1690.

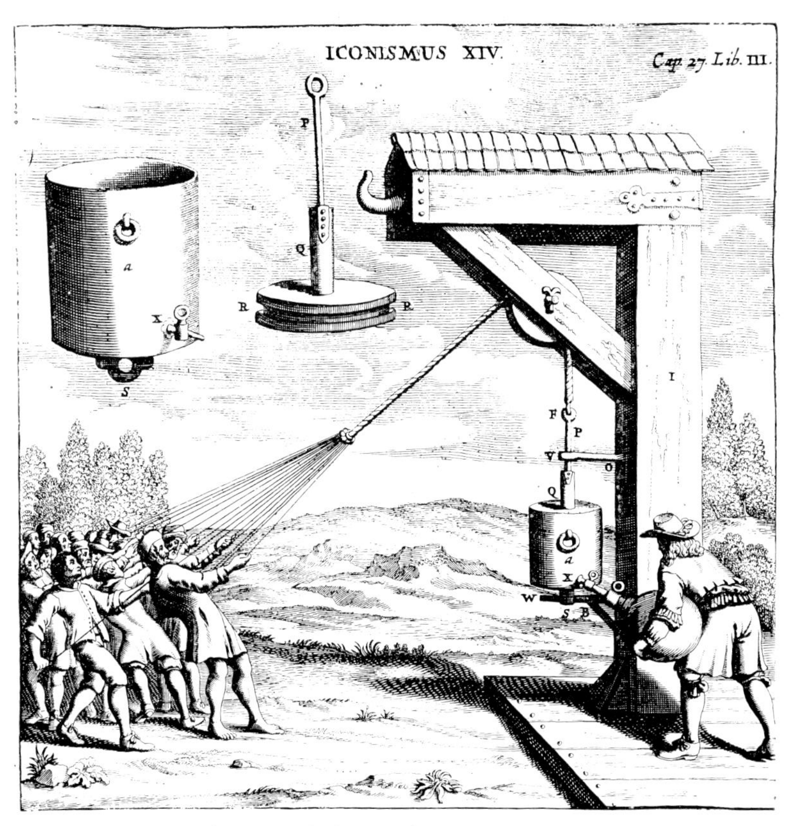

Von Guericke's experimental cylinder and piston 1654.

It was now clear that a useful and powerful engine could be made, if a vacuum could be obtained beneath the piston by some means other than an air pump. About 1679 the celebrated Dutch scientist Christiaan Huygens attempted to expel the air from a cylinder through non-return valves by burning gunpowder, but without success. The difficulties can well be imagined. Huygens' assistant at this time was Denis Papin (1647–1712?), a Frenchman who had studied medicine. Later while he was Professor of Mathematics at Marburg University he continued experimenting on the same lines. In 1690 he took the important step of condensing steam to make the vacuum. He had a small brass cylinder 2½ inches in diameter. When a little water in the bottom was boiled, the steam formed raised the piston to the top, where it was held by a catch. The fire was removed and the steam allowed to condense, producing a partial vacuum. On releasing the catch, the piston descended raising a weight of 60 lbs. Papin suggested a larger machine with several cylinders arranged to turn the paddles of a boat by ratchet work. He did not follow up this suggestion, probably being deterred by the difficulty of making larger cylinders accurately. His model had, however, demonstrated the principles, which were later put to practical use by Newcomen.

Thomas Savery and the First Steam Pump

In 1698 Thomas Savery (1650?–1715) a military engineer and prolific inventor obtained a patent for '... Raiseing of Water... by the Impellent Force of Fire'. The invention was a steam pump on the principles described by Della Porta. Steam from a boiler was admitted into a closed vessel and there condensed by cold water poured on the outside. The resulting vacuum drew water up the suction pipe through a non-return valve in the bottom of the vessel. When steam was again admitted, the water was driven out of the vessel through a second non-return valve and up the delivery pipe. The steam and cold water valves were worked by hand. Savery had great hopes of this invention. He set up a workshop and in 1702 published 'The Miner's Friend' describing an improved pump with two vessels, which discharged alternately to give a continuous flow. As the suction head was limited to about 25 feet and the boiler had to withstand a pressure greater than the delivery head, Savery's pump was only suitable for moderate lifts. It is not known whether it was ever tried in a mine where it would have had to be placed underground. The steam pump installed at York Buildings near the Strand for supplying the neighbourhood with water from the Thames gave endless trouble and was laid aside. Boilers to withstand a pressure of several atmospheres could not be made at that time. However, Savery enjoyed some success with the more modest requirement of lifting water at large private houses.

Late in the 18th century a few of his pumps were used in Manchester in spite of their heavy coal consumption for returning water to the head race of water wheels driving cotton mills. As late as 1820 an improved version with automatic valve gear was at work supplying a waterwheel driving a workshop in London.

Savery was the first to make a useful steam pump and to use a separate boiler, but he chiefly owes his place in the history of the steam engine to his possession of the master patent covering any method of raising water by fire. This patent was extended to 1733.

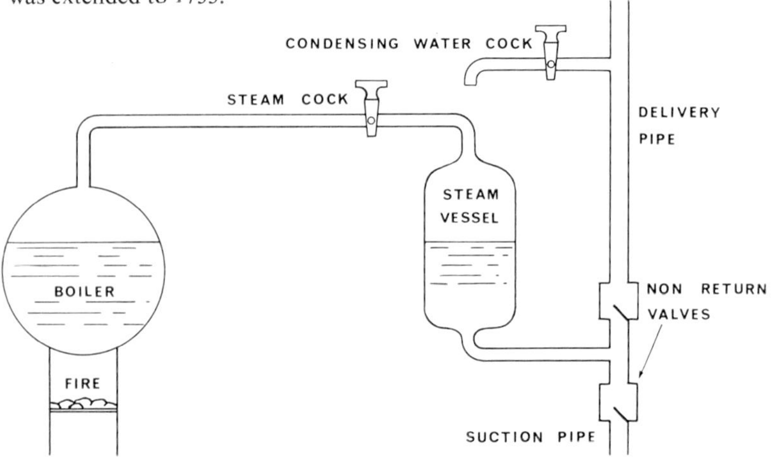

Diagram of Savery's pump.

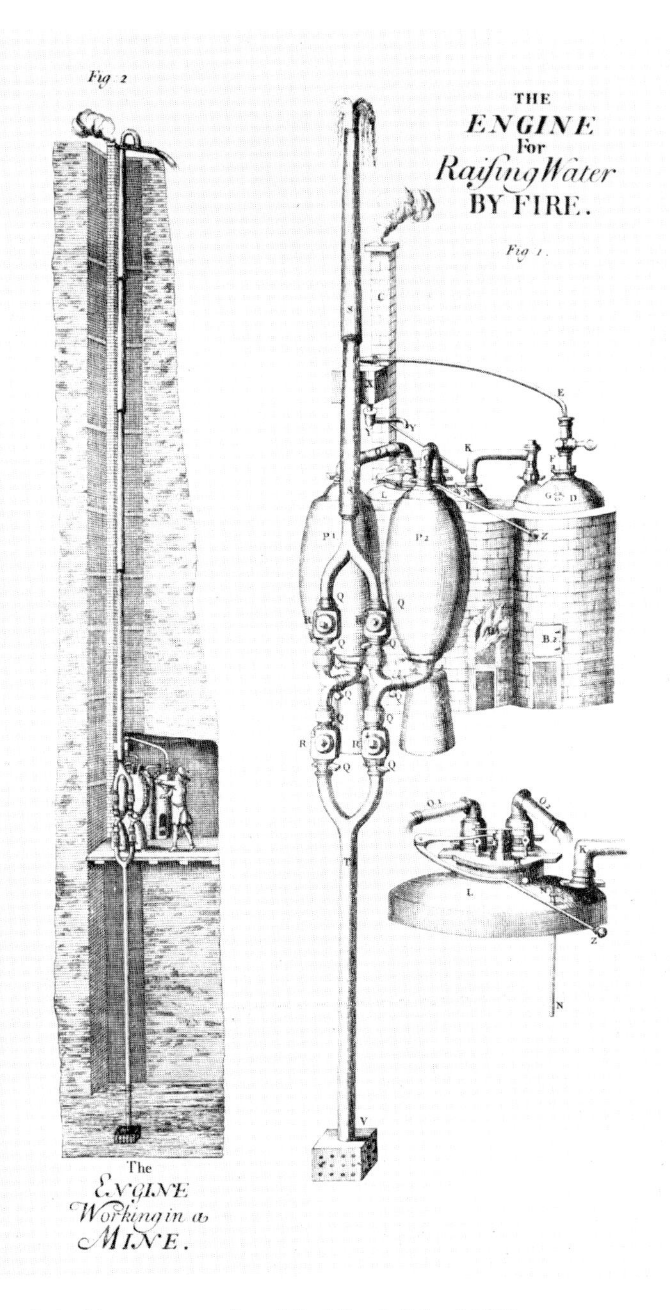

Savery's double steam pump from 'The Miner's Friend' 1702.

Newcomen's Atmospheric Engine

A brief account of Papin's model cylinder and piston experiment of 1690 was published in the Philosophical Transactions of the Royal Society in 1697. It is unlikely that it was seen by Thomas Newcomen (1663–1729) an obscure iron-monger in Dartmouth, but it was probably about this time that he began to work on the same lines. Impelled by the pressing need of the Cornish mines for power for pumping, he pursued his experiments for about fifteen years. However, it was on a colliery near Dudley Castle in Staffordshire in 1712 that he erected the first engine of which there is a definite record. An engraving of 1719 shows this elaborate machine in detail.

Diagram of Newcomen's atmospheric engine 1712.

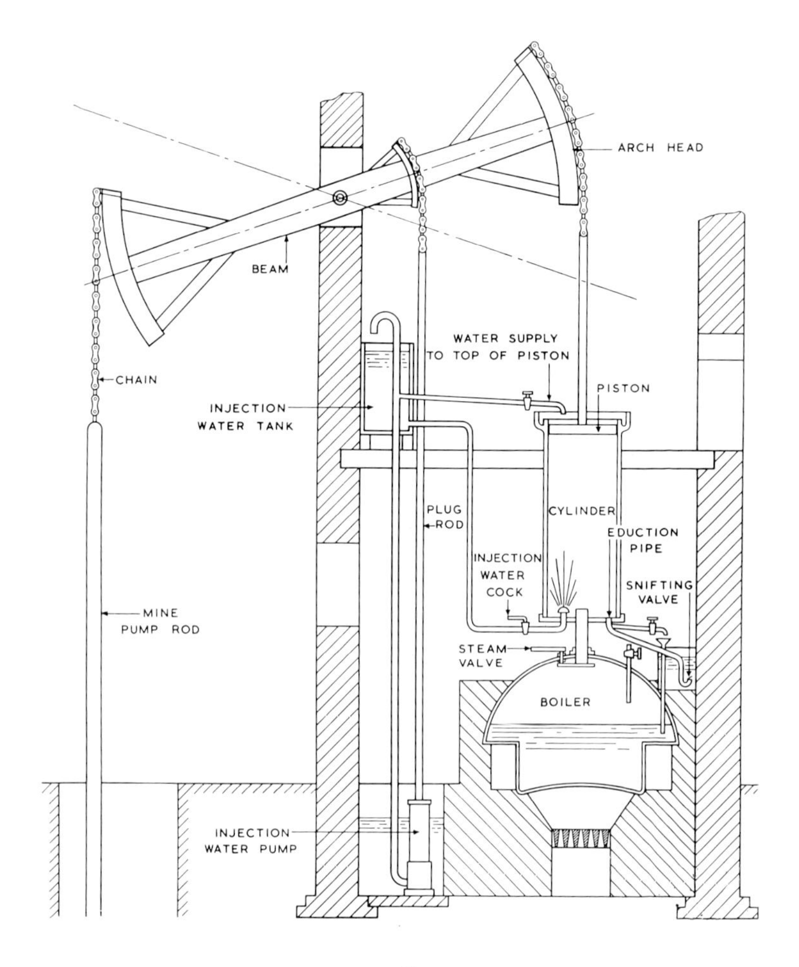

A vertical cylinder open at the top was supplied with steam from a boiler beneath, which resembled a brewer's copper. The piston, packed with leather and sealed with a layer of water on top, was hung by a chain from the arch head of a rocking beam. From the other end of the beam the pump rods were suspended. When steam at slightly above atmospheric pressure was admitted into the cylinder, the piston was drawn up by the weight of the pump rods and any air or water blown out of the cylinder through water sealed non-return valves. After the steam valve was closed, the steam in the cylinder was rapidly condensed by a jet of cold water. The unbalanced atmospheric pressure drove the piston down raising the pump rods and making the working stroke. The cycle was then repeated, the steam valve and injection cock being opened and closed by a plug rod hung from the beam. The Dudley Castle engine had a cylinder of 19 inches internal diameter and made a stroke of about 6 feet. At each stroke it raised 10 gallons of water 51 yards and at 12 strokes per minute developed about 5½ horse power. The design proved thoroughly reliable and remained

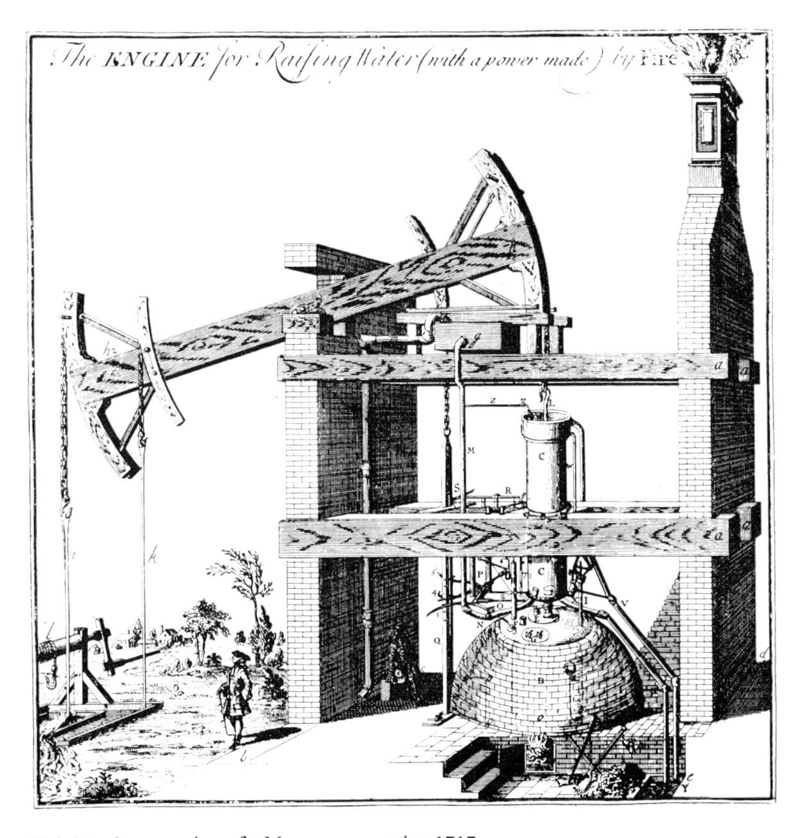

Beighton's engraving of a Newcomen engine 1717.

virtually unaltered for more than half a century, tribute indeed to Newcomen's practical genius.

The invention could not be exploited except by arrangement with Savery, who held the master patent for raising water by fire, the means not being specified. On Savery's death in 1715, the patent rights were acquired by a group of speculators with whom Newcomen seems to have worked amicably enough, though probably in a humbler capacity than he deserved. In spite of the heavy royalty payments demanded, there were engines at work in eight counties by 1716. The first engine on the Continent was put up in 1722 at Königsberg in what is now Czechoslovakia. After the patent expired in 1733 many engines were built, particularly in the colliery districts. By 1775 sixty engines had been erected in Cornwall and there were about one hundred on the Tyne basin alone. The largest of these was at Walker colliery and had a cylinder 74 inches in diameter fed by three boilers with a fourth spare. The engine was also used for public water supply, the first being put up at York Buildings in 1726. A few were returning engines for waterwheels driving machinery or raising coal.

Owing to the very low steam pressure employed – about 1 or 2 lbs per square inch above atmosphere – there was no difficulty in making any part of the engine except the cylinder. At first these were cast in brass and rubbed smooth. Consequently they were very expensive. Soon iron cylinders were being cast and bored at Coalbrookdale at one tenth of the cost. The fuel consumption remained high, because the cylinder had to be heated and cooled at every stroke. About 6 million foot-pounds of useful work were done for each bushel (84 lbs) of coal burnt. From 1772 John Smeaton (1724–1792), the civil engineer, designed a number of engines with improvements in detail, which raised the duty to 9·45 millions, equivalent to a thermal efficiency of less than one per cent. By then the atmospheric engine was obsolescent for James Watt had already invented the separate condenser – the greatest single improvement ever made in the steam engine.

Chelsea Water Works, 1752.

Watt's invention of the Separate Condenser

During the winter of 1763–4 a scale model of an atmospheric engine, belonging to Glasgow University and in need of repair, was placed in the hands of James Watt (1736–1819), instrument maker to the College. Watt was the son of a shipwright at Greenock and had had a year's training in instrument making in London. He was not only a craftsman, but possessed an enquiring and thoughtful mind. His lifelong friend John Robison once said of him 'everything became science in his hands'. So it was that having repaired the model engine and found that it would only make a few strokes before the boiler was exhausted of steam, he began experimenting to find the reason. He measured the volume of steam in comparison with the water from which it was formed and found that at each stroke the model used as much steam as would have filled the cylinder several times. At first he attributed this to heat lost by conduction through the cylinder walls. He therefore constructed another working model with a cylinder made of a non-conducting substance – wood. Further experiments led Watt to realise that the real cause of the waste of steam was the heating and cooling of the cylinder at every stroke. Many years afterwards he wrote:

'I perceived that, in order to make the best use of steam, it was necessary – first, that the cylinder should be maintained always as hot as the steam which entered it; and, secondly, that when the steam was condensed, the water of which it was composed, and the injection itself, should be cooled down to 100°[F], or lower, where that was possible. The means of accomplishing these points did not immediately present themselves; but early in 1765 it occurred to me, that if a communication were opened between a cylinder containing steam, and another vessel which was exhausted of air and other fluids, the steam, as an elastic fluid, would immediately rush into the empty vessel, and continue so to do until it had established an equilibrium; and if that vessel were kept very cool by an injection, or otherwise, more steam would continue to enter until the whole was condensed.'

Thus was the separate condenser invented. Watt proposed to keep the condenser clear of water (from condensation and injection) and air (contained in the steam and from leakage) by means of a pump. He also proposed to keep the cylinder as hot as possible by placing it in a steam jacket surrounded by an insulating substance. For the same reason, steam, instead of atmospheric air, was to press the piston down, the piston rod passing through a stuffing box in the cover.

Watt immediately set to work and made a small model cylinder, condenser and air pump from tin-plate and lead castings. This model has survived, but was only the first of several. At first he tried condensing the steam on the inside of pipes or cells immersed in cold water, but the difficulty of keeping them tight and the risk of their becoming encrusted caused him to revert to condensing by an internal jet. To help meet the cost of these experiments, his friend Dr. Joseph Black, Professor of Chemistry at Glasgow University, made Watt a generous loan and, still more important, introduced him to Dr John Roebuck, an industrial chemist who was working coal mines at Bo'ness on the Firth of Forth. He assumed responsibility for Watt's debt to Black and in return Watt assigned to Roebuck a two-thirds interest in the patent, which he took out in

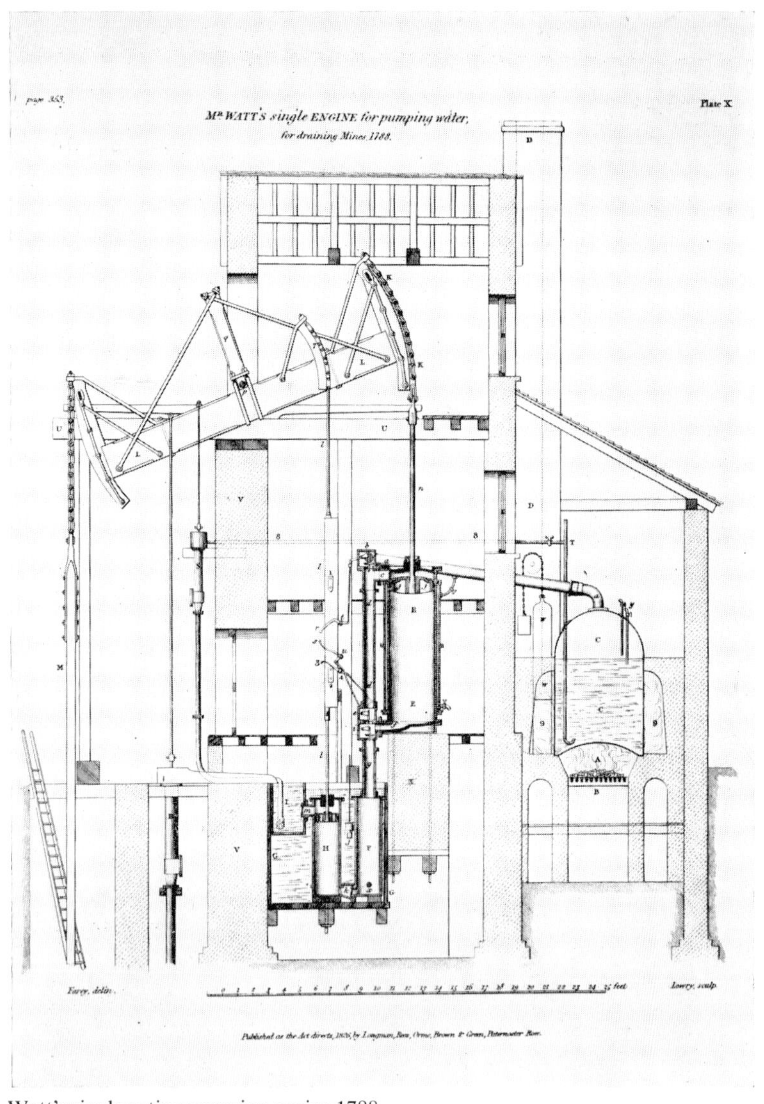

Watt's single acting pumping engine 1788.

1769 for 'A New Method of Lessening the Consumption of Steam and Fuel in Fire Engines'. An experimental engine with a cylinder of 18 inches diameter and 5 feet stroke was set up at Roebuck's house, but Watt was not able to work on it energetically. He had begun to practice as a civil engineer making canal surveys, under the necessity of providing for his family.

In 1773 Roebuck was declared bankrupt. Matthew Boulton of Birmingham, who was one of Roebuck's creditors, had met Watt and had been impressed by his invention. Boulton arranged to take over the two-thirds share in the patent as settlement of the debt. Besides being a leading Birmingham manufacturer of great ability, Boulton had a sanguine temperament and proved to be the ideal partner for Watt. The experimental engine was brought down to Birmingham and Watt was at last able to devote all his time to it. The patent had only eight more years to run, but Watt successfully petitioned Parliament for an extension to 1800. While Watt was still making trials with the experimental engine, Boulton persuaded him to undertake two full size engines, which were completed in 1776. One with a 50 inch cylinder was at Bloomfield Colliery at Tipton in Staffordshire and the other, a 38 inch blowing engine, was for John Wilkinson's blast furnaces at New Willey near Broseley in Shropshire. Wilkinson had just invented a new boring mill, which proved invaluable for making cylinders of the accuracy required for the new engine. No longer could the piston be sealed with water. After many trials Watt hit upon the plan of packing a groove in the periphery of the piston with hemp soaked in tallow. The hemp was compressed by a ring – the junk ring – screwed down upon it. This soft packing proved perfectly satisfactory as long as low pressure steam was used.

The Bloomfield and New Willey engines were an immediate success, consuming less than one third of the fuel of the old engines. Enquiries poured in, particularly from Cornwall, where, within a few years all the atmospheric engines had been replaced. The partners continued the practice whereby the customer paid for all the materials and found the labour for erection. The firm sent drawings and an erector. They also supplied important parts like the valves and valve gear, which were made at Boulton's Soho Manufactory. They insisted that the cylinders were ordered from Wilkinson so that the performance of the engines should not be impaired by bad workmanship. As payment, they claimed one third of the saving in coal over the old atmospheric engines. Two engines on Poldice mine in Cornwall were tested and taken as standard, where no atmospheric engine previously existed. To register the number of strokes from which the premium payments were calculated, a counter worked by a pendulum, was fixed to the engine beam in a locked box, so that it could not be interfered with.

Although the steam pressure remained the same, the new engines were more powerful than the old. They could be loaded to 10½ lbs per square inch of piston instead of only 7 lbs. However they remained single-acting. There were three drop valves with conical seats: steam, equilibrium and exhaust. The steam valve admitted steam above the piston from the boiler and the exhaust valve opened a communication between the space beneath and the condenser, while the piston descended making the working stroke. The equilibrium valve in a pipe connecting the ends of the cylinder opened to allow the piston to ascend on the idle stroke. It soon occurred to Watt that the steam valve could be closed when the piston had only descended through part of its stroke, the steam already in the cylinder expanding and pressing on the piston with diminishing force. In this way less work would be done, but still less steam would be consumed. Watt tried this method of working, but no economy resulted, because the initial pressure was too low. The idea was revived later with higher steam pressures and has assumed great importance. Watt included expansive working in his patent of 1782.

The Rotative Engine

As early as 1769 Watt had devised a rotary engine, or 'Steam Wheel' as he called it, and one was built at Soho in 1774. It consisted of an annular chamber 6 feet in diameter mounted upon a horizontal axis. The chamber was divided into three parts by flap valves at 120° and the lowest part of the chamber was filled with mercury. Steam admitted between the flap valves and the surface of the mercury caused the annular chamber to rotate. It seems to have worked satisfactorily, but was expensive and none was made for sale. It was the first of many attempts to produce rotary motion directly from fluid pressure, a notion which has fascinated inventors. Rotary steam engines have never been a success owing to the difficulty of preventing leakage. They deserve mention, if only to show that the adaptation of the reciprocating engine to produce rotary motion was not an unfortunate consequence of a prior use for pumping, but has a sound practical basis.

Watt would have been content continuing to satisfy the demand for economical pumping engines, but the far-seeing Boulton persuaded him to adapt the reciprocating engine to produce rotary motion. As Boulton wrote in 1781 'the people in London, Manchester and Birmingham are steam mill mad'. The first rotative engine was put up at Soho in 1782 and the first erected outside was supplied to John Wilkinson to drive a hammer. These engines were single acting, and the connecting rod drove the flywheel shaft through a sun-and-planet gear instead of the simpler crank. The use of a crank seems obvious now and was familiar then in the foot lathe. However, neither Watt nor anyone else realised how a flywheel in combination with a crank would regulate the stroke of the piston and carry the crank over the dead centres. While Watt was considering the matter, a crank was applied in 1780 to an atmospheric engine at a mill at Snow Hill, Birmingham, the owner of which, James Pickard, took out a patent, which included the crank. Rather than contest the validity of the patent, Watt devised a number of alternative mechanisms, which he in turn patented in 1782. The sun-and-planet motion was one of these and was quite satisfactory. In fact it had the advantage that it doubled the speed of the flywheel. Sun-and-planet engines were made until 1802, although the firm built engines with a crank before 1794, when Pickard's patent expired.

In 1782 Watt patented the double-acting engine, in which steam acted alternately on both sides of the piston, giving double the power from the same size of cylinder. Making the rotative engine double-acting resulted in a more even motion, but required a rigid connection between the piston rod and the beam. At first in place of the chain, the piston rod was fixed to a toothed rack engaging a toothed sector on the end of the beam. This was noisy and liable to breakdown. Watt therefore connected the piston rod to the beam by a link and devised his parallel motion to guide the top of the piston rod in a straight line. This, the mechanical invention of which he was most proud, was patented in 1784.

In 1788 Watt fitted a conical pendulum governor to control a butterfly throttle valve admitting steam to the engine. This reduced the variation of speed with load and eliminated the risk of the engine running too fast. Watt did not invent the centrifugal governor. It was already in use in windmills for controlling

Watt's rotative beam engine 1788 (*in Science Museum collections*).

the distance between the stones. The double-acting engine needed four valves, two for steam and two for exhaust. In 1799 William Murdock (1754–1839) an employee and later a partner in the firm, patented a sliding valve worked by an eccentric on the crankshaft to take the place of the four valves. The slide valve became standard except for large slow speed engines, which continued to be made with separate valves.

Boulton was correct in his forecast of the demand for rotative engines. They formed more than 60 per cent of the five hundred or so engines built by the firm down to the year 1800. A fixed royalty of £6 in London and £5 in the country per horse power per annum was charged for rotative engines. They were made in standard sizes up to 36 inch cylinder by 7 feet stroke, which at 17 double strokes per minute developed 50 horse power, a unit also devised by Watt.

From the first, the rotative engines were supplied to customers complete. In 1795 a new factory, Soho Foundry, was laid out and equipped solely for building steam engines, so that the firm would be in a strong position when the monopoly ended in 1800.

In the early years of the 19th century new builders of steam engines arose, the foremost being Matthew Murray (1765–1826) of Leeds, whose workmanship was excelled by none. Many improvements in construction followed, iron replacing timber for engine parts. The beam engine remained in favour long after

Model of Maudslay's table engine c. 1815 (*in Science Museum collections*).

direct-acting types had appeared. In a beam engine the working parts are in static equilibrium and the cylinder being vertical cannot become oval through wear. Beam engines were used longest of all for pumping, the last being completed in 1919.

One of the earliest direct-acting types was the table engine developed by Henry Maudslay (1771–1831) from a design he patented in 1807. The vertical cylinder was retained, but drove a crankshaft beneath by return connecting rods. For fifty years it was in favour for such duties as driving workshops.

A direct-acting vertical engine with crankshaft over the cylinder was patented by Phineas Crowther of Newcastle-upon-Tyne in 1800 and, became the standard type of winding engine in that coal field. Many small engines were also made on the same plan. About 1850 James Nasmyth introduced the inverted vertical with the cylinder over the crankshaft and this type has been popular ever since.

The direct-acting horizontal engine first appeared in 1801, but was not regularly manufactured until 1825 when Messrs. Taylor and Martineau began making them for driving sugar mills. Such was the prejudice against horizontal cylinders that this type made little headway until after 1850.

While the new types were sometimes made with a condenser, all used high pressure steam, which forms the subject of the next section.

Model of Horizontal steam engine 1885 (*in Science Museum collections*).

The Introduction of High Pressure Steam

In 1725 Jacob Leupold of Leipzig published a description of a pumping engine worked by steam above atmospheric pressure. After doing work on the piston, the steam was to be discharged into the air.

About a year before he was given the model atmospheric engine for repair, Watt had experimented with a small syringe supplied with steam at several atmospheres pressure. In his patent of 1769 he included a claim for non-condensing engines worked by steam pressure alone, but he never made use of the idea. In fact he was opposed to steam pressures more than a few pounds above the atmosphere, because of the danger of boiler explosions. The wagon boiler he used was quite unsuited to resist higher pressures. New designs were required.

The first engineers to use high pressure steam were Richard Trevithick (1771–1833) in England and Oliver Evans (1755–1819) in America. Trevithick's first high pressure engine was a beam winding engine built in 1800 for Cook's Kitchen mine in Cornwall. The following year he constructed a steam road engine. This had a cylindrical cast iron boiler with an internal return flue of wrought iron. The fire grate was set in one end of the flue and the chimney at the other. The cylinder was vertical and sunk into the boiler to keep it hot. Return connecting rods drove the rear wheels from a crosshead on the piston rod. The steam was distributed by a four-way cock and discharged up the chimney to increase the draught. The engine ran successfully and Trevithick used the same design for a number of self-contained portable engines, which were built for rolling iron, driving sugar mills and similar purposes. Unfortunately one of his boilers exploded at Greenwich in 1803 owing to the safety valve being fastened down. Misrepresentation of the cause of the accident did much to prevent the wider use of these engines.

In 1802 he had an experimental pumping engine built, which worked with steam at 145 lbs per square inch, but 50 lbs or less was the usual limit for his engines. In 1804 a locomotive based on the portable engine design, but with a horizontal cylinder and both axles driven by gearing, successfully hauled a load of 25 tons on the Penydarren tramway in South Wales. Trevithick designed a similar locomotive for a colliery near Newcastle-upon-Tyne, which probably caused George Stephenson to take up the steam locomotive with such far-reaching results.

Meanwhile, Oliver Evans in America was building high pressure engines. He used cylindrical boilers, at first with an internal furnace flue like Trevithick's, but latterly with an external fire beneath. The steam pressure was 100 lbs. per square inch or more, but Evans was handicapped, because mechanical engineering in America was not then as far advanced as in England. The first commercially successful steamboat in America, Robert Fulton's 'Clermont' of 1807, had a Boulton and Watt low pressure engine. Evans campaigned for ten years before his high pressure engines were used for this purpose. After his death in 1819 many steamboats were built with high pressure engines for the great rivers of the interior.

Trevithick's high pressure portable engine c. 1805 (*in Science Museum collections*).

The Cornish Pumping Engine

In 1806 Trevithick proposed to replace the wagon boilers of a Boulton and Watt pumping engine on Dolcoath mine by his cylindrical boilers and to work the engine with steam at 25 lbs per square inch. The steam valve was to be shut early in the stroke, so that the remainder of the stroke would be made by expansion of the steam already in the cylinder, just as Watt had tried with low pressure steam. Nothing was done at the time, but in 1812 Trevithick had the opportunity to try out his ideas on a little pumping engine, which he put up on Wheal Prosper mine at Gwithian. This was a single acting condensing engine with a cylinder 24 inches in diameter and a stroke of 6 feet. It was supplied with steam at 40 lbs per square inch by a cylindrical boiler 6 feet in diameter and 24 feet in length with an internal furnace tube running from end to end – a type of boiler, which became known as a 'Cornish' boiler. At first when the load was light, the steam was cut off at between one ninth and one tenth of the stroke. This was the first 'Cornish' engine, but its development was left to others, as Trevithik sailed for South America soon afterwards.

In 1811 Captain Joel Lean began to publish Monthly Reports of the duty performed by the pumping engines in Cornwall. These reports excited such a competitive spirit amongst the engineers that their influence towards improvement can hardly be overestimated. After 1800, when Boulton and Watt severed their connection with Cornwall, the duty of the pumping engines declined. In 1813 the best engine only did 26·4 million foot lbs per bushel (94 lbs) of coal, while the average for the 29 engines reported was 19·5 millions. By lagging all the exposed surfaces of the boilers, cylinder and steam pipes Samuel Grose attained a duty of 76·8 millions with an 80 inch engine on Wheal Towan in 1828. In the same year the average for the 57 engines reported was 37·1 millions.

In 1834 William West's 80 inch engine at Fowey Consols did 125 millions in a twenty-four hour trial. As this engine usually did less than 100 millions, doubt has been cast on the result of the trial and it is said that the preparations for it included stacking firewood in the boiler flues! However, there was no denying the accuracy of the Reports published by the Leans. The results were so far in advance of the performance of the low pressure engines used elsewhere that they could not be ignored. Thomas Wicksteed purchased a duplicate of Fowey Consols engine and had it erected at the East London Water Works, Old Ford, in 1838. It met its guarantee duty of 90 millions, which was 2¼ times that of the best Boulton and Watt low pressure engine. Cornish engines became the standard plant for Water Works for thirty years. They were retained even longer on the Cornish mines, the last being re-erected in 1924. Thanks to the efforts of the Cornish Engines Preservation Society, several engines remain in the Duchy. When the Cornish engine became well-known and the reasons for its high duty understood, it greatly influenced steam engineering practice. Rotative condensing engines were built to work expansively on high pressure steam. However, in the Cornish engine the expansion of steam in a single cylinder reached its practical limit. The way to further advance was by the compound engine.

Model of Taylor's Cornish pumping engine at Gwennap United Mines 1840. At one period it regularly performed a duty of over 100 million foot lbs per bushel of coal. The cylinder was 85 inches in diameter and the stroke 11 feet. (*in Science Museum collections*).

The Compound Engine

In 1781 Jonathan Hornblower (1753–1815) patented an engine with two cylinders in which the steam acted in turn, the second cylinder being larger than the first. It was found to be no more economical than the single cylinder engine and, as it had a separate condenser and infringed Watt's patent, few were built.

In 1804 Arthur Woolf (1776–1837) revived the idea, but with high pressure steam and took out a patent for it. He erected his first engine at Meux's Brewery in London, where he was employed and supplied it with steam from a cast iron tubular boiler, which he had also invented. Unfortunately he had an erroneous idea of the law of expansion of steam and made the high pressure cylinder far too small, so that the engine would not do the work required. As a result he left Meux's employ and, in partnership with Humphrey Edwards, built a number of compound engines with the cylinders more correctly proportioned. These were quite successful and showed a saving of about 50 per cent compared with a Boulton and Watt low pressure engine. Edwards later migrated to France where he obtained a patent in 1815, and put up about three hundred Woolf engines, but their manufacture was continued by other firms in London. Meanwhile in 1811 Woolf returned to his native Cornwall and erected several large compound pumping engines. They were quite successful at first, but were ultimately found to do no better than the single cylinder engines, so few compounds were built.

In 1845 most large factories and mills were still driven by low pressure beam engines which were not sufficiently strongly constructed to be worked expansively with high pressure steam on the Cornish system. William McNaught had the idea of adding a smaller high pressure cylinder on the crank side of the beam between its centre and the connecting rod. This avoided overstressing the beam. He then put in new high pressure boilers and worked the engine as a

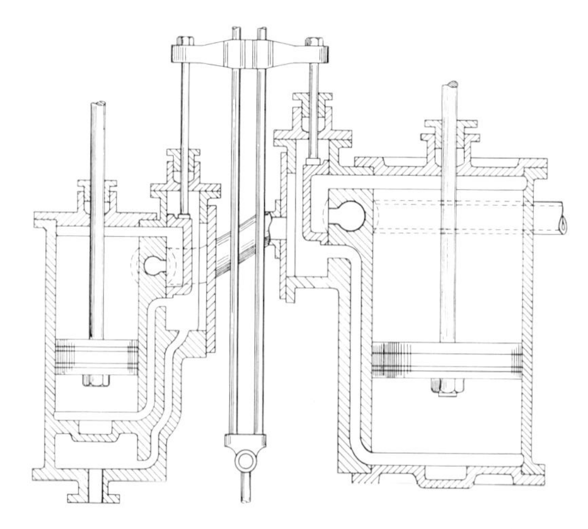

Section through the cylinders of a compound beam engine.

Compound beam engine c. 1860 (*in Science Museum collections*).

compound with an increase in both power and economy. New compound beam engines were built with the cylinders side by side and compounding spread to other types of engine.

Compounding enabled the steam to be expanded in the engine to many times its original volume with less loss from condensation on the cylinder walls. It also reduced leakage losses and the stresses in the engine. Eventually about 1900 when steam pressures reached 150 lbs per square inch on land, the expansion was carried out in three cylinders – triple expansion.

The Corliss Engine

In America in 1850 George Henry Corliss (1817–1888) brought out an engine with rocking cylindrical valves, four to each cylinder. The advantage was thermodynamic, for the inlet valves and ports were not cooled by the exhaust steam. Also, a horizontal cylinder could be very effectively drained by placing the exhaust valves at the bottom. The valves were each linked to a wrist plate, which was oscillated by an eccentric, The steam valves were released by a trip gear controlled by the governor and closed smartly under the action of springs. This arrangement gave a very close speed control, which was valuable for textile mill drives, but it also prevented the engines being run at speeds higher than 80–100 revolutions per minute.

Corliss engines were very economical and eventually compound and triple expansion engines of several thousand horse power were built. They were very little known in Europe until the Paris Exposition of 1867 after which several firms in Great Britain began to make them. The type probably reached the zenith of its popularity about 1900. By this time superheating had been introduced. It was found that heating the steam to a temperature above that of the water from which it was formed resulted in a marked economy by reducing condensation in the cylinder. Corliss valves were only suitable for steam, which was moderately superheated. For high superheat drop valves were essential. These were of the double seated equilibrium type developed in Cornwall in the early years of the nineteenth century to enable the large valves on the pumping engines to be opened easily against high pressure steam.

Model of Horizontal engine with Corliss valves 1898 (*in Science Museum collections*).

High Speed Steam Engines

About 1880 a demand arose for engines for dynamo driving. At first slow speed engines drove dynamos by belt, but the advantages of direct drive led to the development of high speed steam engines. The latter were made single acting, so that there was no reversal of load on the crankpins. The object was to avoid knocking, which developed unless the bearing clearance was reduced to a point at which seizing was inevitable. Most of these single acting designs were short lived, but in 1884 and 1885 Peter William Willans (1851–1892) patented his central valve engine, which held its own for twenty years. In this design the steam was distributed by a piston valve inside a hollow piston rod, worked by an eccentric on the crankpin. The engine was vertical, totally enclosed and splash lubricated. It was made with tandem cylinders as simple, compound or triple expansion and with one, two or three cranks. It was manufactured to close limits so that the parts were interchangeable. It was ultimately built in sizes up to 2,500 horse power and was very economical.

The Willans engine was rendered obsolete by the steam turbine about 1904, but before then it had a formidable rival in the double-acting engine built by Belliss and Morcom of Birmingham using forced lubrication on the system patented by their chief draughtsman Albert Charles Paine in 1890 and 1892.

Willans two crank compound central valve engine and dynamo c.1888. The engine developed 18 horse power at 450 revolutions per minute.

(in Science Museum collections).

A valveless oscillating pump forced oil into the main bearings, whence it was conveyed by passages in the crankshaft to the crankpins and by pipes to the small end bearings and crosshead slippers. The eccentric rods of the piston valves were similarly lubricated. At first, these engines were not very economical, but by careful design and the use of triple expansion, the steam consumption was halved and rivalled that of the much more expensive and complex slow speed engine. Triple expansion Belliss engines were built in sizes up to 2,900 horse power.

Section through a two crank compound Willans central valve engine.

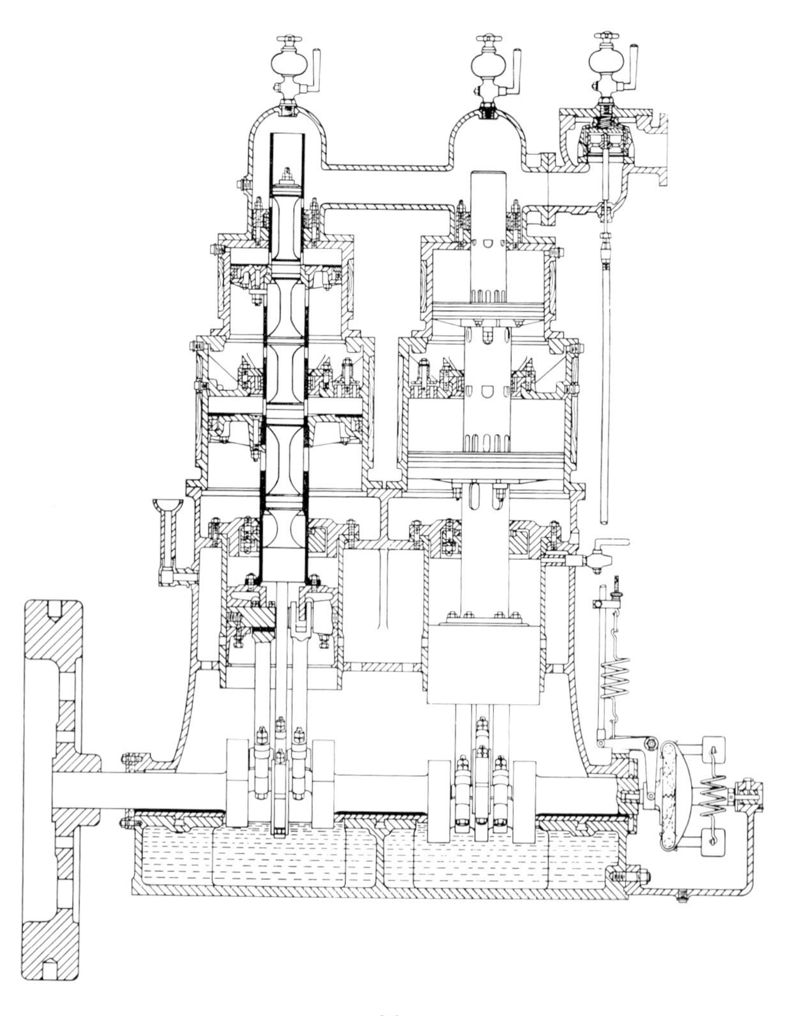

Like the Willans engine, the Belliss was superseded by the steam turbine for central power stations, but the enclosed forced lubricated design was taken up by other manufacturers and has been built in great numbers. They are used in factories, hospitals and laundries, where low pressure steam is required anyway for space heating or process work. By generating the steam at a higher pressure and passing it first through an engine, electric power can be produced at very small additional cost.

Section through a Belliss and Morcom compound self-lubricating engine.

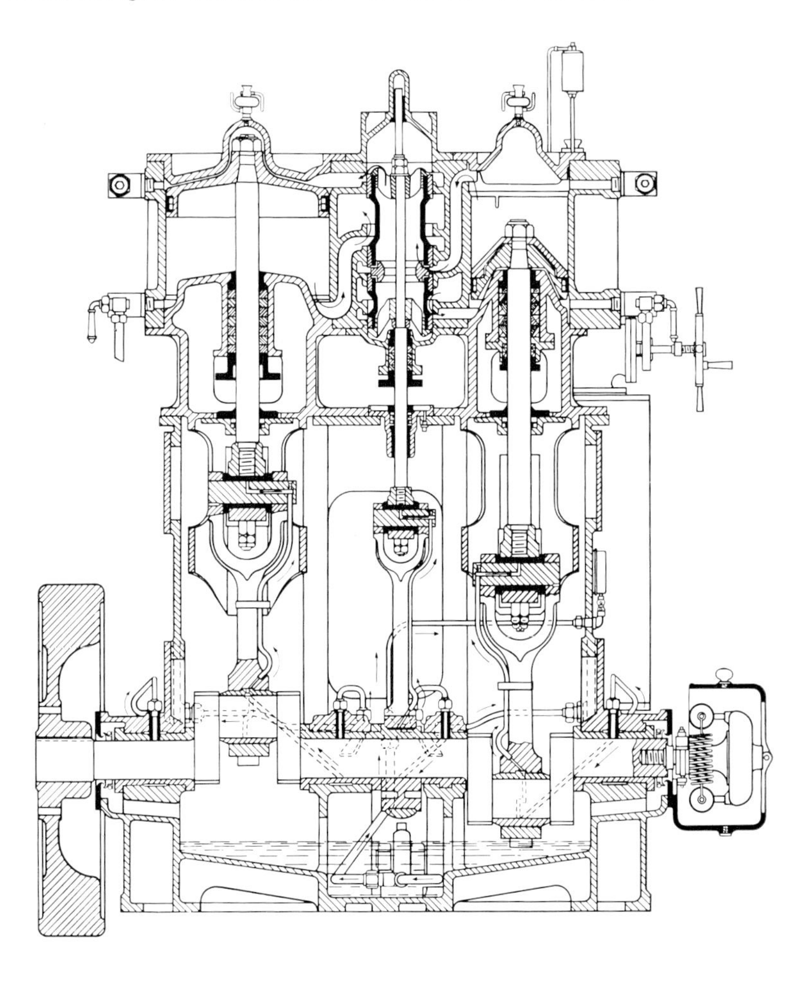

The Uniflow Engine

In 1827 Jacob Perkins (1766–1849) invented and built an experimental engine working at the very high pressure of 1400 lbs per square inch. The engine was single acting and instead of an exhaust valve, ports in the cylinder were uncovered by the piston at the end of its stroke. In 1881 Leonard Jennett Todd patented a double-acting engine with a trunk piston, which uncovered a ring of exhaust ports in the centre of the cylinder in a similar way. To quote Todd's own words, his object was 'to produce a double-acting steam engine which shall work more efficiently, which shall produce and maintain within itself an improved gradation of temperature extending from each of the two Hot Inlets to its common central Cold Outlet, which shall cause less condensation of the entering steam'. However the constructional difficulties were not overcome until 1908 when Professor Johann Stumpf of Charlottenburg re-introduced the engine using drop type inlet valves. It was tried for many purposes including locomotives, but achieved success as a condensing stationary engine. It was as economical as a triple expansion engine and occupied less space. Several British firms took out licences and built it for about twenty years.

Progress in Fuel Economy

Almost all the improvements in the steam engine have been directed towards reducing the consumption of fuel. Improvements in boilers have also played their part. The boiler passes most of the heat in the fuel to the engine in the steam, but it degrades the heat so that the engine can only convert a small fraction of it into mechanical work. Raising the pressure of the steam and hence its temperature improves this fraction. However the earliest engines were so wasteful that there was scope for improvement without raising the steam pressure. The progress made in two centuries is illustrated by the following table for engines doing similar work, namely pumping.

Approximate Coal Consumption per water horse power per hour

		lb
1712	Newcomen atmospheric engine	32
1772	Atmospheric engine improved by Smeaton	17
1776	Watt engine with separate condenser	9
1834	Cornish engine	3
1870	Horizontal condensing compound engine	2
1885	Vertical condensing triple expansion engine	1½

Model of Robey Uniflow engine (*in Science Museum collections*).

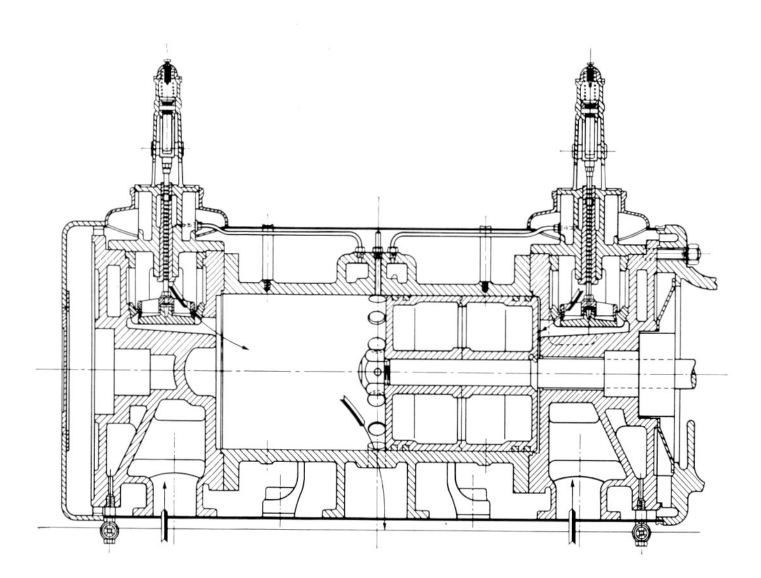

Section through the cylinder of a Robey Uniflow engine.

The Development of Land Boilers

Hitherto little has been said of boilers. The haystack boiler used with the atmospheric engines was probably derived from a brewer's copper. It had a concave bottom with the fire grate set beneath. On their way to the chimney the hot gases were carried in a brick flue round the sides of the boiler. This was known as 'wheel draught'. The earliest haystack boilers were made of copper with a lead top, but after 1725 hammered wrought iron plates took the place of the copper.

The wagon boiler used by Watt had a cross section similar to the haystack, but the elongated form gave more heating surface and was easier to make. The boiler was set like the haystack with wheel draught, but the larger wagon boilers

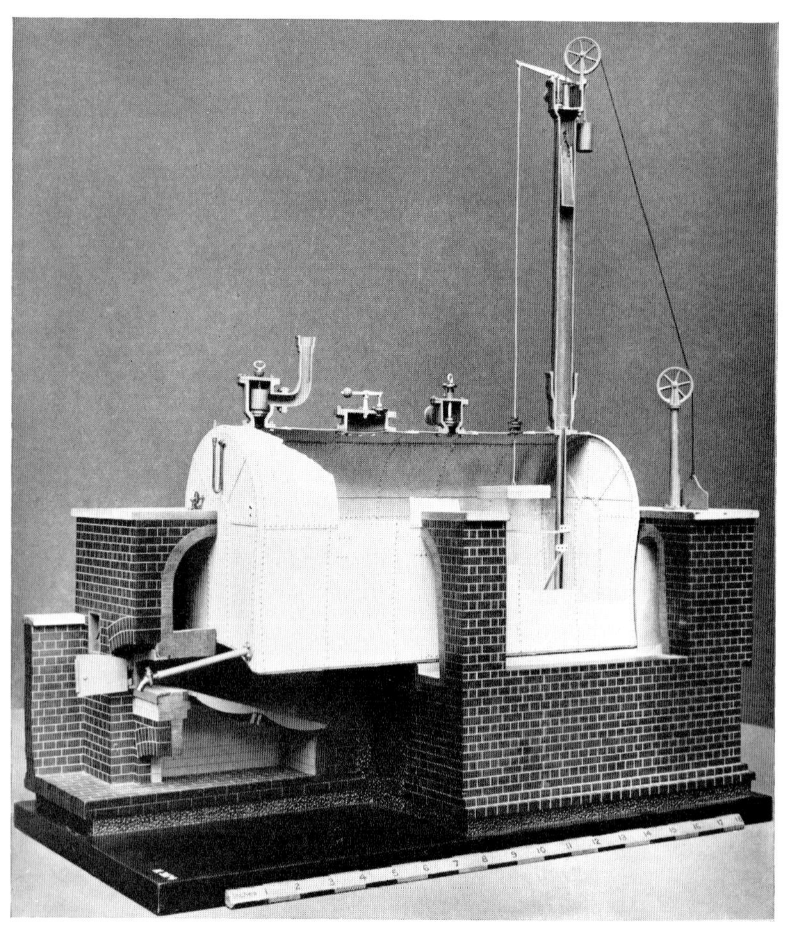

Model of Wagon boiler as made in the early part of the nineteenth century (*in Science Museum collections*).

Model of Babcock and Willcox boiler c. 1903 with superheater above the water tubes *(in Science Museum collections).*

often had in addition a more or less rectangular flue running through them from end to end. Narrow rolled iron plates became available about 1795 and replaced the small hammered plates.

In Trevithick's high pressure Cornish boiler the grate was set in one end of an internal fire tube. At the rear end the gases divided and passed forward in brick flues along each side before returning beneath the bottom to the chimney. In 1844 William Fairbairn and John Hetherington of Manchester patented what became known as the Lancashire boiler, although there is evidence of its use in Cornwall ten years earlier. The Lancashire boiler resembled the Cornish, but had two furnace tubes. The patentees claimed that by firing the furnaces alternately the smoke would be consumed. The Lancashire boiler was easier to clean and had a larger grate area without the weakness of a single large internal flue. It gradually supplanted the Cornish boiler. Mild steel was permitted for boilers in place of wrought iron in 1865. Improvements in detail enabled the Lancashire boiler to withstand higher pressures. In 1904 John Thompson of Wolverhampton introduced dished ends, which eliminated internal stays and were able to withstand pressures of 250 lbs per square inch and more. When superheaters were introduced about 1890 the tubular elements were placed in the flue leading to the chimney.

Woolf's cast iron high pressure boiler of 1804 has already been mentioned. A horizontal drum was connected by branches to a number of horizontal tubes placed transversely beneath it. The tubes and the lower half of the drum were heated by the fire and hot gases which circulated past them. The tubes were provided with a removable cover at one end so that they could be cleaned, but the circulation was defective and cast iron unsuitable. Woolf's boiler was only one of many attempts to increase heating surface by means of tubes, but no success could be achieved until better materials were available.

A successful design of water tube boiler was patented in America in 1867 by Herman Babcock and Stephen Willcox and was made for many years. It had an inclined bank of straight tubes connected at each end to a drum above in which the steam and water separated. The tubes, being straight, were easy to clean and the circulation was excellent.

Another successful type of water tube boiler was that patented by A. Stirling, also in America, in 1889. This had steeply inclined tubes expanded into three steam-and-water drums at their upper ends and below into two mud drums in which sediment was deposited. The water tube boiler has since been developed in larger sizes and for higher pressures to meet the needs of the steam turbine.

Model of Stirling's water tube boiler 1889 (*in Science Museum collections*).

EXPONENTIAL BUSINESS GROWTH

THE POWER OF A PRIVATE BANKING LICENSE

TAIMOUR ZAMAN

Private banking is a concept that has been around for centuries, providing tailored financial services to high-net-worth individuals and families. From its origins in medieval Italy to its modern-day form, private banking has played a vital role in the global economy. But what exactly is private banking? And how has it evolved?

In this book, I delve deep into private banking, exploring its history, significance, and relevance in the modern era. From the licensing process and the challenges faced by applicants to compliance and regulations, wealth management, investment banking, and popular destinations for private banks, we cover a wide range of topics that will help you understand the ins and outs of this fascinating industry.

But I don't stop there. I also explore private banks' challenges in day-to-day operations, including regulatory compliance, customer expectations, and technology disruptions. I provide real-world examples of how to overcome these challenges and achieve success.

I also look into the future of private banking and examine the emerging trends that will shape the industry from my perspective. From digitalization to changing customer expectations, I explore the factors driving change in the industry and what it means for private banks and their clients.

At the heart of this book is the belief that private banking is about building relationships and delivering personalized solutions that meet each client's unique needs. It's about providing exceptional service, expert advice, and innovative solutions that help clients achieve their financial goals.

So, whether you're a private banker looking to stay on top of industry trends, an entrepreneur or business owner looking to leverage private banking services to grow your business, or a high-net-worth individual looking for tailored financial solutions, this book is for you. Join me on a journey through the fascinating world of private banking and discover the opportunities and challenges ahead.

Chapter Roadmap

In this chapter, I introduce the concept of private banking, its history, and its significance. I discuss how private banking has evolved from its origins in medieval Italy to its modern-day form, where it plays a vital role in the global economy. I also explore the differences between private banking and retail banking.

In this chapter, I dive into the licensing process for private banks. I discuss the various steps involved, the requirements for obtaining a license, and the challenges applicants face. I provide insights on navigating the process successfully and offer tips on what to avoid.

This chapter discusses the opportunities and challenges of obtaining a private bank license. I analyze the current market trends and provide an overview of the competition. I also examine the different private bank types and their business models.

In this chapter, I discuss private banks' challenges in day-to-day operations. I explore the regulatory environment, customer expectations, technology disruptions, and other factors affecting a private bank's performance. I provide examples of how to overcome these challenges and achieve success.

In this chapter, I focus on the importance of compliance and regulations for private banks. I explore the laws and regulations that private banks must adhere to and provide an overview of the compliance requirements. I also provide some tips on how to maintain compliance effectively.

This chapter discusses the relationship between wealth management and private banking. I explore wealth management strategies and how private banks can help clients achieve their financial goals. I provide case studies of successful wealth management strategies for your review and consideration.

I invite you to investigate the relationship between investment banking and private banking in this chapter. I will show you the different investment banking services that private banks offer and provide an overview of the investment banking landscape. I also provide case studies of successful investment banking strategies.

In this chapter, I will discuss the opportunities that private banking offers successful entrepreneurs and business owners. I will show you the different services private banks offer to help entrepreneurs and business owners achieve their financial goals. I provide case studies of successful private banking strategies for successful entrepreneurs and business owners.

This chapter will show you the costs of acquiring private bank licenses in different geographies. I will show you the different licensing requirements and the costs of obtaining a license in different regions. I will provide an overview of the licensing fees and other costs of operating a private bank in different geographies.

In this chapter, I will discuss the ideal business models for private banks. I will explore the different types of private banks and their business models. I will provide an overview of the different services private banks can offer and the clients they can serve.

In this chapter, I discuss the popular destinations for private banks and why they are popular. I explore the factors that make specific regions attractive to private banks, including the regulatory environment, tax laws, and business opportunities. I provide an overview of the most popular destinations for private banks.

In this chapter, I will provide you with possible future trends in private banking and explore the factors driving change in the industry, including digitalization, changing customer expectations, and regulatory developments. In this chapter, I want to provide insights into the future of private banking and what it may look like in the coming years. I discuss emerging trends, such as artificial intelligence and automation, the growing demand for sustainable investments, and the increasing importance of cybersecurity. I will also provide recommendations on how private banks can adapt and thrive in these changes.

CHAPTER 1

Introduction to Private Banking and Its Significance

Private banking has a long and storied history, dating back to medieval Italy, where wealthy families needed a safe place to store their wealth and valuables. Over the centuries, private banking has evolved into a sophisticated industry vital to the global economy.

Today, private banking is defined as personalized services offered to high-net-worth individuals (HNWIs) and ultra-high-net-worth individuals (UHNWIs) with complex financial needs that require customized solutions. Private banking services include wealth management, investment banking, estate planning, and trust services.

Private banking is distinct from retail banking, which provides standardized banking services to individuals and small businesses. Private banking clients typically have investable assets of at least $1 million, while retail banking customers may have only a few thousand dollars in their accounts.

Private banking has grown in importance over the years as globalization and increasing wealth has created a need for sophisticated financial services. Private banks now play a critical role in managing the wealth of HNWIs and UHNWIs, helping them to preserve and grow their wealth over the long term.

Private banking has also evolved in response to changing customer needs and technological advancements. Digitalization has transformed how private banks deliver their services, allowing clients to access their accounts and manage their finances anywhere in the world.

To understand the importance of private banking, consider the analogy of a captain and a ship. The captain is responsible for navigating the ship, ensuring it stays on course and avoids potential hazards. Private bankers similarly guide their clients through the complex financial landscape, helping them to avoid potential risks and achieve their financial goals.

One example of a private bank that has successfully adapted to changing customer needs and technological advancements is JPMorgan Chase. The bank has invested heavily in its digital platform, allowing clients to access their accounts and manage their finances on their smartphones or computers. JPMorgan Chase has expanded its services to include alternative investments, such as private equity and hedge funds, providing clients with a more diversified investment portfolio.

In conclusion, private banking has come a long way since its origins in medieval Italy. Today, it plays a critical role in the global economy, providing customized financial solutions to HNWIs and UHNWIs. Private banking has evolved in response to changing customer needs and technological advancements and will continue to do so.

CHAPTER 2

The Licensing Process for Private Banks

Obtaining a private bank license can be daunting for any aspiring banker. However, it is possible to navigate the process successfully with careful planning and preparation. In this chapter, we will dive into the licensing process for private banks, discussing the various steps involved, the requirements for obtaining a license, and the challenges that applicants face. We will provide insights on navigating the process successfully and offer tips on what to avoid.

The first step in the licensing process is identifying the jurisdiction where you want to establish your private bank. Each jurisdiction has a regulatory framework, and the license requirements vary significantly. It is essential to research and understands the regulatory environment in the jurisdiction of interest, as this will inform the licensing strategy and approach.

Once you have identified the jurisdiction, the next step is to prepare a comprehensive business plan. The business plan should detail the proposed activities of the private bank, the target market, the organizational structure, and the financial projections. The business plan is a crucial document informing the regulator of the bank's intentions, capabilities, and financial stability.

After preparing the business plan, the next step is to prepare the application documents. The application documents typically include the business plan, financial statements, compliance policies and procedures, anti-money laundering policies and procedures, and other relevant documentation. The application documents should be comprehensive and well-prepared, as the regulator will thoroughly review them.

Once the application documents are complete, they should be submitted to the regulator, along with the application fee. The regulator will then review the application documents and assess the applicant's suitability. The assessment process can be lengthy, and the regulator may require additional information or clarification.

During the assessment process, the regulator may conduct an on-site inspection of the bank's premises and operations. The inspection is intended to ensure that the bank follows the regulatory requirements and has adequate systems and controls to manage the risks associated with private banking.

After completing the assessment process, the regulator will decide whether to grant the license. If the license is granted, the bank can then commence operations. However, the regulatory oversight does not end with the granting of the license, and the bank will be subject to ongoing monitoring and supervision by the regulator.

Navigating the licensing process for private banks can be challenging and complex. However, with careful planning and preparation, it is possible to navigate the process successfully. One key to success is engaging with experienced professionals, such as lawyers and consultants, who can provide guidance and support throughout the process.

Case Study: Bank of Singapore

The Bank of Singapore is a leading private bank in Asia and the Middle East. The bank was established in 2010 and has increased thanks to its focus on delivering personalized wealth management solutions to high-net-worth individuals.

One of the keys to the Bank of Singapore's success is its licensing strategy. The bank has pursued a multi-jurisdictional licensing approach, obtaining licenses in several key markets, including Singapore, Hong Kong, and Dubai. This approach has allowed the bank to serve clients in multiple markets and expand its reach while mitigating regulatory risk.

The Bank of Singapore's licensing strategy was informed by careful research and planning and a deep understanding of the regulatory environment in each jurisdiction. The bank engaged with experienced professionals, such as lawyers and consultants, to navigate the licensing process successfully.

The Bank of Singapore's licensing strategy is an excellent example of how careful planning and preparation can lead to success in the licensing process for private banks. By pursuing a multi-jurisdictional approach and engaging with experienced professionals, the bank could quickly navigate the complex and challenging licensing process.

The Bank of Singapore's licensing strategy focused on obtaining licenses in several key jurisdictions worldwide, including Singapore, Hong Kong, and Dubai. By diversifying its license portfolio, the bank expanded its global reach and offered its clients a more comprehensive range of services.

One of the critical factors in the Bank of Singapore's success was its commitment to thorough preparation and due diligence. The bank engaged with experienced professionals in each jurisdiction to ensure a comprehensive understanding of the regulatory requirements and licensing procedures.

Additionally, the bank was careful to ensure that it met all the requirements for each license application, including financial stability, compliance with anti-money laundering regulations, and adequate risk management procedures.

As a result of its careful planning and preparation, the Bank of Singapore successfully obtained licenses in multiple jurisdictions, allowing it to provide its clients with a wide range of services and establish itself as a leading private bank in the region.

CHAPTER 3

Opportunities & Challenges in Obtaining a Private Bank License

Private banking has become a lucrative business for those with the resources and expertise to enter the market. However, obtaining a private bank license is a challenging feat. The process is long and complicated, with many challenges along the way. Despite these challenges, the rewards of a successful private banking business can be substantial. This chapter will explore the opportunities and challenges of obtaining a private bank license and provide insights on navigating them successfully.

Opportunities

The private banking industry is increasing, presenting numerous opportunities for those seeking entry. According to a report by Research and Markets, the global private banking industry is expected to grow at a compound annual growth rate of 6.7% between 2020 and 2025. This growth is driven by the increasing number of high-net-worth individuals (HNWIs)

Worldwide, the Asia-Pacific region is seeing the most significant growth in HNWIs.

The demand for private banking services increases as clients seek tailored financial solutions. Private banks can offer wealth management, estate planning, tax advisory, and investment banking services. These services are highly profitable, with private banks earning significant client fees.

Furthermore, the competition in the private banking industry is relatively low compared to the retail banking sector. This provides an excellent opportunity for new players to enter the market and establish themselves as reputable private banks.

Challenges

Despite the opportunities the private banking industry presents, obtaining a private bank license can take time and effort. The regulatory environment is complex, with varying requirements depending on the jurisdiction. Additionally, the licensing process can be lengthy, with many requirements, such as minimum capital requirements and compliance with anti-money laundering regulations.

Another challenge in the private banking industry is the intense competition. Private banks have established themselves as reputable institutions and built strong relationships with their clients. New entrants must overcome this barrier by offering differentiated services and building their brand reputation.

Different Types of Private Banks and Their Business Models

There are different types of private banks, each with a unique business model. For instance, family offices serve high-net-worth families and their businesses, providing various services such as wealth management, estate planning, and tax advisory. On the other hand, boutique private banks specialize in serving ultra-high-net-worth individuals, providing highly personalized services.

Large private banks, such as UBS and Credit Suisse, offer various financial services, including wealth management, investment banking, and asset management. However, smaller private banks, such as EFG International, may focus on a particular niche, such as serving clients in emerging markets.

Case Study: Lombard Odier

Lombard Odier is a Swiss private bank that has been in operation since 1796. The bank's business model focuses on providing highly personalized services to its clients, offering tailored solutions to meet their financial needs.

One of the bank's unique features is its partnership structure. The bank is wholly owned by its partners and its clients. This structure aligns the bank's interests with its clients, creating a culture of client service and satisfaction.

Lombard Odier has successfully navigated the licensing process for private banks, expanding its operations to 26 locations worldwide. The bank has established itself as a reputable institution known for its personalized services and innovative solutions.

Conclusion

Obtaining a private bank license can be challenging, but the rewards of a successful private banking business can be substantial. The opportunities presented by the growing private banking industry are significant, and the low competition in the sector provides new players an excellent opportunity to establish themselves as reputable institutions.

To navigate the challenges of obtaining a private bank license successfully, it is essential to understand the current market trends and the competitive landscape. As we have seen, there are different types of private banks, each with unique business models and target clients. Aspiring private bankers need to identify their niche and develop a compelling value proposition that caters to the needs of their target clients.

The licensing process for private banks is complex, and applicants need to be well-prepared and engage with experienced professionals to navigate the process successfully. Compliance and regulatory requirements are crucial, and private banks must maintain high levels of compliance to ensure their ongoing operation.

Finally, private banks must be adaptable and innovative to stay competitive in the fast-evolving private banking industry. With technology advancements, changing customer expectations, and regulatory developments, private banks must stay abreast of the latest trends and continually improve their services to stay ahead of the competition.

In summary, obtaining a private bank license is challenging but rewarding. With careful planning, preparation, and persistence, aspiring private bankers can establish themselves as reputable institutions in the growing private banking industry. The opportunities are endless, and the sky's the limit for those willing to take the leap of faith and pursue their private banking dreams.

CHAPTER 4

Challenges in Running a Private Bank

Private banks face various daily operations challenges impacting their performance and success. This chapter will examine some of these challenges and explore strategies for overcoming them.

Regulatory Environment

One of the biggest challenges facing private banks today is the regulatory environment. Governments worldwide are introducing more stringent regulations to combat money laundering and other financial crimes. These regulations can be complex and expensive, making it difficult for smaller private banks to compete with larger institutions. Private banks must invest in compliance resources and stay up-to-date with regulatory changes to overcome this challenge.

Customer Expectations

Another challenge for private banks is meeting the changing expectations of customers. Today's clients expect a personalized experience, a wide range of investment options, and a high level of service. Private banks must invest in technology to provide clients with the services they demand, including online portals and mobile apps that make it easy to access account information and complete transactions.

Technology Disruptions:

The rapid pace of technological change can also present challenges for private banks. Advancements in artificial intelligence, machine learning, and automation are transforming the financial services industry, and private banks must adapt to these changes to remain competitive. Private banks must invest in new technologies that allow them to automate repetitive tasks and free up time for their employees to focus on more valuable activities, such as client engagement and relationship management.

Other Factors

Private banks face market volatility, changing economic conditions, and geopolitical risks. These factors can impact the performance of a private bank's investment portfolios and make it difficult to manage risk effectively. Private banks must stay vigilant and adapt their investment strategies to changing market conditions to meet their client's investment objectives.

In conclusion, private banks face various challenges in their day-to-day operations. However, private banks can overcome these challenges by investing in compliance resources, technology, and human capital and delivering a superior client experience. Private banks must remain agile and adaptable, continually adjusting their strategies to meet the changing needs of their clients and the financial marketplace.

CHAPTER 5

Compliance and Regulations for Private Banks

Private banks operate in a highly regulated environment, and compliance with laws and regulations is crucial for maintaining their reputation and credibility. The compliance landscape for private banks is constantly evolving, and private banks must remain up to date with the latest developments to avoid regulatory penalties and reputational damage.

The regulatory framework for private banks varies from country to country. Some critical regulation areas that private banks must comply with include anti-money laundering (AML) regulations, Know Your Customer (KYC) requirements, tax regulations, data privacy laws, and investment and securities regulations.

AML regulations require private banks to have robust systems in place to prevent money laundering and the financing of terrorism. KYC requirements demand that private banks verify the identities of their customers and assess the potential risks of doing business with them. Tax regulations require private banks to comply with local and international tax laws and report suspicious transactions to the authorities.

Data privacy laws are another crucial area of regulation for private banks. Customers entrust private banks with sensitive financial and personal information, and private banks must safeguard this information from unauthorized access or disclosure.

Private banks must also comply with investment and securities regulations that vary by jurisdiction. These regulations govern how private banks offer and manage investment products, ensuring they are transparent, fair, and in line with customer expectations.

The compliance requirements for private banks can be overwhelming, but maintaining compliance is critical to ensure the longevity and success of a private banking business. Private banks must have a robust compliance framework with clear policies and procedures, effective training programs, and regular internal and external audits.

One of the most significant challenges private banks face in maintaining compliance is keeping up with the evolving regulatory landscape. Regulations can change quickly, and private banks must remain alert to any new developments that could affect their compliance obligations. Regular engagement with regulatory authorities and industry bodies is essential to stay current on the latest developments.

Another challenge for private banks is managing customer expectations around compliance. Customers expect private banks to have rigorous compliance standards and to protect their assets from financial crime. Private banks must communicate compliance policies and procedures to customers and ensure a transparent and open dialogue around compliance-related issues.

In conclusion, compliance and regulations are of utmost importance to private banks, and they play a critical role in maintaining the reputation and credibility of the private banking industry. Private banks must remain vigilant and proactive in complying with regulatory requirements and have a robust compliance framework. Compliance is essential for regulatory compliance, building customer trust, mitigating risks, and creating a sustainable business model.

CHAPTER 6

The Relationship Between Wealth Management and Private Banking

Wealth management is a complex and often misunderstood term for creating, protecting, and growing wealth. It involves financial planning, investment management, and other strategies that help individuals and families achieve their financial goals. Private banking, on the other hand, refers to the provision of banking and investment services to high-net-worth individuals and families.

The relationship between wealth management and private banking is critical. Private banks offer wealth management services that help clients achieve their financial objectives. This chapter will explore wealth management strategies and how private banks can help clients achieve their financial goals.

Understanding Wealth Management

Wealth management is not just about investing money; it involves a range of financial strategies designed to help individuals and families achieve their long-term financial goals. Wealth management strategies typically include:

Financial Planning involves a comprehensive plan considering a client's financial goals, risk tolerance, and time horizon. The plan typically includes retirement planning, tax planning, and estate planning.

Investment Management: This involves managing a client's investment portfolio to maximize returns while minimizing risk. Investment management strategies typically include diversification, asset allocation, and active management.

Risk Management involves identifying and mitigating potential risks that could impact a client's wealth. Risk management strategies typically include insurance and asset protection.

Estate Planning involves creating a plan to transfer a client's wealth to their heirs or beneficiaries. Estate planning strategies typically include wills, trusts, and other estate planning vehicles.

Private Banking and Wealth Management

Private banks offer wealth management services to high-net-worth individuals and families. These services typically include:

Investment Advisory Services: Private banks advise clients and manage their investment portfolios. They offer a range of investment products, including stocks, bonds, mutual funds, and alternative investments.

Financial Planning: Private banks offer comprehensive financial planning services considering a client's long-term financial goals and risk tolerance.

Estate Planning: Private banks provide services to help clients transfer their wealth to their heirs or beneficiaries.

Credit and Lending Services: Private banks offer credit and lending services to their clients, including personal loans, mortgages, and lines of credit.

The Importance of Private Banking in Wealth Management

Private banking is an essential component of wealth management. High-net-worth individuals and families require specialized services not typically available through retail banking channels. Private banks offer a range of products and services tailored to the needs of their clients, including customized investment solutions, comprehensive financial planning, and specialized credit and lending services.

Case Study: UBS Wealth Management

UBS is a global wealth management firm that provides financial services to high-net-worth individuals and families. The firm offers customized investment solutions, comprehensive financial planning, and specialized credit and lending services.

UBS's wealth management strategy is focused on delivering value to its clients. The firm provides innovative products and services to help its clients achieve long-term financial goals. For example, UBS offers a range of investment products, including exchange-traded funds (ETFs), alternative investments, and structured products.

Conclusion

Wealth management and private banking are closely related, and private banks play a critical role in helping high-net-worth individuals and families achieve their

financial goals. Private banks offer a range of products and services tailored to the needs of their clients, including customized investment solutions, comprehensive financial planning, and specialized credit and lending services. The relationship between wealth management and private banking is critical, and private banks are well-positioned to help their clients achieve their long-term financial objectives.

CHAPTER 7

Investment Banking and Private Banking: A Dynamic Relationship

The world of finance is vast and complex, with many different sectors and specializations. Among them, the most important is an investment banking and private banking. While these two sectors may seem distinct, they are closely intertwined, with private banks often relying on investment banking services to offer their clients a broader range of products and services.

This chapter will explore the relationship between investment banking and private banking. We will look at the various investment banking services that private banks offer, how private banks work with investment banks, and the benefits that this relationship can bring for clients and institutions.

Investment Banking Services Offered by Private Banks

Private banks offer a range of investment banking services to their clients. These services include underwriting securities, advisory services, mergers and acquisitions, and securities trading. Private banks also offer services related to capital markets, including equity and debt issuance, initial public offerings (IPOs), and bond offerings.

One of the key benefits of working with a private bank for investment banking services is that the bank has a deep understanding of the client's financial situation and investment goals. This allows the bank to offer personalized investment advice and tailor investment strategies to the client's needs.

The Investment Banking Landscape

The investment banking landscape is constantly evolving, with new players entering the market and new technologies emerging. Today, investment banks face various challenges, including increased competition, tighter regulations, and the growing importance of technology.

Despite these challenges, investment banking remains a crucial part of the financial sector, providing critical services to clients and helping to drive economic growth.

Working Together: Private Banks and Investment Banks

Private and investment banks often work together to offer a range of services to their clients. This collaboration can take many forms, from co-underwriting securities to joint advisory services.

One of the key benefits of this collaboration is that private banks can leverage investment banks' expertise and resources to offer their clients a broader range of investment options. For example, a private bank may work with an investment bank to offer a new bond offering to its clients or to provide advisory services on a potential merger or acquisition.

Case Studies: Successful Investment Banking Strategies

One example of a successful investment banking strategy by a private bank is offering IPOs to clients. By underwriting IPOs, private banks can offer their clients access to new investment opportunities and the potential for significant returns. For example, in 2019, Swiss private bank UBS underwrote the IPO of ride-hailing giant Uber, allowing its clients to invest in one of the most highly anticipated IPOs of the year.

Another example of a successful investment banking strategy is offering advisory services to clients on mergers and acquisitions. Private banks can leverage their deep understanding of their client's financial situations and goals to offer personalized advice on potential mergers and acquisitions, helping clients to make informed decisions and navigate complex transactions.

Conclusion

In conclusion, the relationship between investment and private banking is dynamic and constantly evolving. Private banks offer their clients investment banking services, including underwriting, advisory services, and securities trading. By working with investment banks, private banks can leverage the expertise and resources of these institutions to offer a broader range of investment options and services to their clients.

As the investment banking landscape evolves, private banks must remain agile and adaptable, working to stay ahead of new trends and technologies while maintaining a deep understanding of their client's needs and goals. With the right strategies in place, private banks can continue to offer high-quality investment banking services and help their clients to achieve their financial goals.

CHAPTER 8

Private Banking for Entrepreneurs and Business Owners

Entrepreneurs and business owners face unique challenges when managing their wealth. They often have complex financial situations, including income streams, investments, and tax considerations. Private banking can offer tailored solutions to help entrepreneurs and business owners manage their wealth effectively and achieve their financial goals.

Private banks offer entrepreneurs and business owners various services, including investment management, tax planning, and cash management. Private bankers work closely with their clients to understand their financial needs and goals and create personalized strategies to help them achieve success.

One of the most significant benefits of private banking for entrepreneurs and business owners is access to investment opportunities that may be limited to the public. Private banks have extensive networks and connections that can provide access to exclusive investment opportunities, such as private equity and venture capital investments.

Private banks can also guide tax planning and optimization, which is particularly important for entrepreneurs and business owners. They can help clients navigate complex tax laws and regulations, manage their tax liabilities, and take advantage of tax-saving strategies.

Case Study: Private Banking for a Successful Entrepreneur

John is a successful entrepreneur who founded a technology startup that was acquired for a substantial amount. He received significant money from the acquisition and needed a trusted advisor to help him manage his wealth. John turned to a private bank for assistance.

The private bank worked closely with John to understand his financial goals and developed a personalized wealth management strategy. They provided investment management services, helping John to allocate his assets across a range of investments, including private equity and venture capital investments.

The private bank also provided tax planning and optimization services, helping John to minimize his tax liabilities and take advantage of tax-saving opportunities. Additionally, the bank helped John with his philanthropic efforts by setting up a charitable foundation and managing his donations.

Thanks to the tailored solutions provided by the private bank, John grew his wealth while achieving his philanthropic goals.

Analogies: Private Banking for Entrepreneurs and Business Owners is like a trusted co-pilot that helps navigate the complexities of managing wealth, providing personalized solutions to help clients reach their financial destination.

Private banking can also be compared to a tailor who creates custom-fit solutions designed to meet each entrepreneur's or business owner's unique needs, providing a perfect fit for their financial goals and objectives.

In conclusion, private banking offers entrepreneurs and business owners unique opportunities to manage their wealth effectively and achieve their financial goals. With access to exclusive investment opportunities, personalized solutions, and expert advice, private banking can help entrepreneurs and business owners take their wealth to new heights.

CHAPTER 9

Cost of Acquiring a Private Bank License in Different Geographies

Private banking is a highly regulated industry, and obtaining a license to operate a private bank can be costly. The costs associated with acquiring a private bank license vary depending on the geography and jurisdiction in which the bank is established. This chapter will explore the licensing requirements and costs of obtaining a private bank license in different regions.

Asia-Pacific

In the Asia-Pacific region, private banking is a growing industry with a high demand for wealth management services. However, the regulatory requirements for private banking are strict, and the cost of acquiring a license can be high. In Singapore, for example, the cost of a private bank license can range from SGD 250,000 to SGD 1,500,000, depending on the size and complexity of the bank's operations. In Hong Kong, the cost of a private bank license can range from HKD 2,000,000 to HKD 5,000,000.

Europe

In Europe, private banking is a well-established industry with a solid regulatory framework. The cost of acquiring a private bank license in Europe can vary widely depending on the jurisdiction in which the bank is established. In Switzerland, for example, the cost of a private bank license can range from CHF 50,000 to CHF 500,000, depending on the size and complexity of the bank's operations. In the United Kingdom, the cost of a private bank license can range from £50,000 to £250,000.

Americas

In the Americas, private banking is a growing industry with a strong demand for wealth management services. The regulatory requirements for private banking can be strict, and the cost of acquiring a license can be high. In the United States, for example, the cost of a private bank license can range from $500,000 to $1,500,000, depending on the size and complexity of the bank's operations. In Canada, the cost of a private bank license can range from CAD 250,000 to CAD 1,000,000.

Factors Affecting the Cost of Acquiring a Private Bank License

Several factors can affect the cost of acquiring a private bank license, including the regulatory environment, the size and complexity of the bank's operations, and the jurisdiction in which the bank is established. Banks operating in highly regulated jurisdictions typically face higher licensing fees and compliance costs. Banks with more complex operations, such as those with international or many clients, may also face higher licensing fees and compliance costs.

Conclusion

Acquiring a private bank license can be costly, with the cost varying widely depending on the jurisdiction in which the bank is established. However, the cost of acquiring a license is only one factor to consider when starting a private bank. Banks must also consider the regulatory environment, the size and complexity of their operations, and the target market when planning to establish a private bank. With careful planning and preparation, banks can navigate the licensing process successfully and establish themselves as reputable institutions in the private banking industry.

CHAPTER 10

Ideal Business Models for Private Banks

Private banks come in different shapes and sizes, each with its unique business model. While there is no one-size-fits-all approach to private banking, successful private banks tend to have specific characteristics that distinguish them from their peers. This chapter will explore the ideal business models for private banks, including their services and clients.

Types of Private Banks

There are two main types of private banks: independent and affiliated. Independent private banks are standalone institutions that offer a wide range of financial services, including wealth management, investment banking, and lending. On the other hand, Affiliated private banks are part of a larger financial institution and typically offer a narrower range of services.

In addition to these two types of private banks, boutique private banks specialize in a particular niche, such as sustainable investing or philanthropic advising. These boutique banks have a smaller client base but can be highly specialized and offer unique services.

Business Models for Private Banks

Successful private banks have a business model focusing on client needs and delivering exceptional service. Private banks must be able to offer a wide range of financial services to their clients, including wealth management, investment banking, lending, and estate planning. They must also be able to provide customized solutions that meet each client's specific needs.

A critical aspect of a successful private bank's business model is its ability to build long-term client relationships. Private banks that can establish trust and maintain strong relationships with their clients are more successful than transactional ones.

Another critical aspect of a successful private bank's business model is its ability to leverage technology to deliver a superior client experience. Private banks that can provide clients with easy-to-use digital platforms and advanced data analytics tend to be more successful than those that rely solely on traditional methods.

Types of Clients

Private banks typically serve high-net-worth individuals, families, and business owners. These clients have complex financial needs and require specialized services that retail banks do not typically offer. Private banks must be able to offer a range of services, including wealth management, investment banking, lending, and estate planning, that can help clients achieve their financial goals.

Private banks must also be able to provide customized solutions that meet each client's unique needs. This requires a deep understanding of each client's financial situation, risk tolerance, and investment objectives.

Conclusion

In conclusion, successful private banks have a business model focused on delivering exceptional service and building long-term client relationships. Private banks must be able to offer a wide range of financial services, leverage technology to deliver a superior client experience, and provide customized solutions that meet the unique needs of each client. By focusing on these critical elements, private banks can establish themselves as trusted advisors and build a strong reputation in the industry.

CHAPTER 11

Popular Destinations for Private Banks and Why

The private banking industry has experienced significant growth in recent years, and as a result, private banks are always on the lookout for new destinations to set up their operations. The decision to establish a private bank in a particular region depends on several factors, including the regulatory environment, tax laws, business opportunities, and overall economic conditions. In this chapter, we will explore some of the most popular destinations for private banks and why they are ideal for establishing a private banking business.

Switzerland: The Home of Private Banking

Regarding private banking, Switzerland is undoubtedly the most popular destination for private banks. Switzerland has a long history of banking secrecy, and its regulatory environment is highly conducive to establishing private banks. Swiss banks are renowned for their discretion, reliability, and expertise in managing clients' assets. Additionally, the country has a stable political and economic environment, making it an attractive destination for private banks.

Hong Kong: Gateway to Asia

Hong Kong is another popular destination for private banks. The city is regarded as a gateway to Asia, and its strategic location makes it an attractive destination for private banks looking to expand their operations in the region. Hong Kong has a robust regulatory environment and a stable political and economic environment, making it an ideal location for private banks.

Singapore: A Rising Star

Singapore has emerged as a popular destination for private banks in recent years. The city-state has a highly developed financial industry, and its regulatory environment is highly conducive to private banking. Singapore has also established itself as a hub for wealth management services, making it an attractive destination for private banks looking to serve high-net-worth clients in the Asia-Pacific region.

Dubai: The Hub of the Middle East

Dubai is rapidly emerging as a hub for private banking in the Middle East. The city has established itself as a hub for international business, and its regulatory environment is highly conducive to private banking. Dubai's location also makes it an attractive destination for private banks looking to serve clients in the Middle East and North Africa region.

The Bahamas: A Tropical Paradise

The Bahamas is a popular destination for private banks looking to serve clients in the Caribbean and Latin America. The country has a highly developed financial industry and a regulatory environment conducive to private banking. Additionally, the Bahamas' tropical climate and natural beauty make it an attractive destination for clients looking to enjoy their wealth in a luxurious environment.

Conclusion

In conclusion, private banks have several options for choosing a location to set up their operations. The choice of destination depends on several factors, including the regulatory environment, tax laws, business opportunities, and overall economic conditions. As private banking continues to grow, we expect to see new destinations emerge as popular destinations for private banks. Private banks should choose a destination that best suits their business model and clients' needs.

CHAPTER 12

Future Trends in Private Banking

The private banking industry has evolved rapidly in recent years, driven by technological advancements, shifting customer expectations, and changing regulatory requirements. As we look to the future, it's clear that the industry will continue to change and evolve, creating new opportunities and challenges for private banks.

Digitalization: The Rise of Fintech

The rise of fintech has been one of the most significant trends in the financial services industry in recent years, and private banking is no exception. Fintech companies have been disrupting traditional business models by offering innovative products and services that leverage new technologies like AI and blockchain.

To stay competitive, private banks must embrace digitalization and invest in new technologies to enhance their offerings and improve the customer experience. This could include developing mobile apps, leveraging AI and automation to streamline processes, and using data analytics to understand client needs and preferences better.

Changing Customer Expectations: Personalization and Sustainability

Customers' expectations are evolving, and private banks must adapt to meet these changing needs. Clients now expect a more personalized approach to wealth management, with tailored solutions that meet their needs and preferences.

In addition, the demand for sustainable investing is growing, and private banks must incorporate environmental, social, and governance (ESG) factors into their investment strategies to remain relevant.

Regulatory Developments: Stricter Compliance and Data Protection

The regulatory environment is becoming increasingly complex, with stricter compliance requirements and more significant.

Emphasis on data protection. Private banks must comply with all relevant laws and regulations, including anti-money laundering and know-your-customer requirements.

In addition, data protection is becoming a more significant concern, and private banks must take steps to safeguard client information and prevent data breaches.

The Emergence of New Markets: Asia and Africa

As traditional markets become more competitive, private banks seek to expand into new regions, including Asia and Africa. These markets present significant opportunities for growth but also pose unique challenges, including cultural differences and regulatory requirements.

To succeed in these markets, private banks must invest in local talent and build strong relationships with regulators and key stakeholders.

Recommendations: Embrace Change and Innovate

To thrive in the future, private banks must embrace change and innovate. This includes investing in new technologies, developing personalized solutions for clients, and

We are incorporating sustainable investing principles into investment strategies.

In addition, private banks must focus on building solid relationships with clients and adapting to changing regulatory requirements. By doing so, they can position themselves as leaders in the industry and continue to grow and evolve in the years to come.

Conclusion

The future of private banking is bright, but it will require banks to adapt and innovate to remain relevant in a rapidly changing environment. By embracing new technologies, incorporating ESG factors into investment strategies, and building solid client relationships, private banks can position themselves for success in future years.

Acknowledgments

I am overwhelmed with gratitude and would like to take a moment to express my heartfelt thanks to my creator for blessing me with this beautiful life and the opportunity to make a difference in this world. I am also profoundly grateful to my parents, especially my mother, for their unwavering love and support throughout my life. And to my beloved wife, who has been my rock through thick and thin, I owe you a debt of gratitude that I can never fully repay. Your unwavering support and encouragement have driven my success, and I will be forever grateful. Thank you all from the bottom of my heart.

I also want to thank and acknowledge ChatGPT for enhancing the copy and helping me with the book title.

Acknowledgments for sources of data

For Chapter 1, titled: Introduction to Private Banking and its History, I want to acknowledge the sources of data:

- Historical documents, books, and research on the origins and evolution of private banking
- Industry reports and publications on the role and significance of private banking in the global economy
- Interviews with industry experts and practitioners on the history and development of private banking

For Chapter 2, titled: The Licensing Process for Private Banks, I want to acknowledge the sources of data:

- The relevant regulatory bodies or government agencies issue regulations and guidelines.
- Interviews with regulatory officials or experts in the licensing process
- Case studies and experiences shared by successful private bank license holders.

For Chapter 3, titled: Opportunities and Challenges in Getting a Private Bank License, I want to acknowledge the sources of data:

- Industry reports and market research on the current state and

trends of the private banking industry
- Interviews with industry experts, analysts, and practitioners on the challenges and opportunities of obtaining a private bank license.
- Case studies and experiences shared by successful private bank license holders.

For Chapter 4, titled: Challenges in Running a Private Bank, I want to acknowledge the sources of data:

- The relevant regulatory bodies or government agencies issue regulations and guidelines.
- Industry reports and publications on private banks' challenges in their day-to-day operations.
- Interviews with industry experts, analysts, and practitioners on private banks' regulatory, technological, and customer-related challenges.

For Chapter 5, titled: Compliance and Regulations for Private Banks, I want to acknowledge the sources of data:

- Laws, regulations, and guidelines issued by the relevant regulatory bodies or government agencies.
- Industry reports and publications on compliance and regulatory requirements for private banks
- Interviews with compliance experts or regulatory officials

For Chapter 6, titled: Wealth Management and Private Banking, I want to acknowledge the sources of data:

- Industry reports and research on wealth management strategies and practices
- Case studies and experiences shared by successful private banks in their wealth management services.
- Interviews with wealth management experts and practitioners.

For Chapter 7, titled: Investment Banking and Private Banking, I want to acknowledge the sources of data:

- Industry reports and publications on investment banking services and practices
- Case studies and experiences shared by successful private banks in their investment banking services.
- Interviews with investment banking experts and practitioners.

For Chapter 8, Private Banking for Entrepreneurs and Business Owners, I want to acknowledge the sources of data:

- Industry reports and research on entrepreneurs' and business owner's financial needs and preferences.
- Case studies and experiences shared by successful private banks in their services for entrepreneurs and business owners.
- Interviews with business experts and practitioners.

For Chapter 9, Cost of Acquiring a Private Bank License in Different Geographies, I want to acknowledge the sources of data:

- Regulations and guidelines are issued by the relevant regulatory bodies or government agencies in different regions.
- Industry reports and publications on the costs of obtaining a private bank license in different regions.
- Interviews with industry experts, analysts, and practitioners on the costs and requirements of obtaining a private bank license in different regions.

For Chapter 10, Ideal Business Models for Private Banks, I want to acknowledge the sources of data:

- Industry reports and research on private banks' different business models and services.
- Case studies and experiences shared by successful private banks

in their business models.
- Interviews with industry experts, analysts, and practitioners on the ideal business models for private banks.

For Chapter 11, Popular Destinations for Private Banks and Why, I want to acknowledge the sources of data:

- Industry reports and publications on the most popular destinations for private banks.
- Research and analysis of the regulatory environment, tax laws, and business opportunities in different regions.
- Interviews with industry experts, analysts, and practitioners on the factors that make specific regions attractive to private banks.

For Chapter 12, titled: Future Trends in Private Banking, I want to acknowledge the sources of data:

- Industry reports and publications on the latest trends and developments in the private banking industry.
- Research and analysis of emerging technologies and their impact on the industry.
- Interviews with industry experts, analysts, and practitioners on the future of private banking and the strategies for adapting to changes.

Please note: Financial Services (just like any other industry) go through rapid change. For up-to-date content, please subscribe to our YouTube Channel at: http://bit.ly/3Zrt1a3

About AltFunds Global

AltFunds Global helps successful businesses leverage their assets (Cash, Standby letters of credit, land, etc.) to access capital that banks offer to their top 30 clients.

The company specializes in structured finance projects and private placement programs.

You can contact AltFunds Global through its website: www.altfundslgobal.com

About The Author:
Taimour Zaman - Founder and chief capital strategist of AltFunds Global.

Hi, I'm Taimour Zaman.

As a child, I was fascinated with spies and superheroes like James Bond and Batman. Combining technology and ingenuity to save the world resonated with me.

Naturally, after graduating from university, I gravitated towards big tech. I spent over a decade consulting with giants like Adobe, Avaya, Microsoft, SAP, IBM, and TATA Consulting Service. That experience earned me invitations to serve as a keynote speaker at various universities and teach at the University of Ontario Institute of Technology.

Concerned about Canada's declining innovation rankings, in 2012, I co-founded a global non-profit called One Million Acts of Innovation with a straightforward intention: to enhance Canada's innovation capabilities by connecting arts and youth to business. The goal was to count one million acts of innovation by unleashing creative ideas that businesses could use to create meaningful change in the world.

That inspired co-founding of an investment firm called 8 Billion Acts of Innovation. The firm funded artificial intelligence and blockchain companies on a television show that put aside $20 million for investment purposes.

Fast forward to today, I am now the Principal of Alt Funds Global. My firm provides capital to various industries, monetizes funding instruments, and assists high-net-worth individuals in private placement programs and other important investment decisions.

I am the author of Hyper-Growth: How to Connect to Customers in New Ways, Structured Finance Demystified, How to Profit from the Shifting Currents in Global Markets: Currency Trading and Intermarket Analysis, Clearing and Settlement: A Practical Guide, and my YouTube channel on monetizing financial instruments has garnered thousands of views.

Openly individualistic, I often jokingly credit my success to not following my father's financial advice. However, I did inherit my father's love of family. My wife and I have an equally individualistic and strong-minded young daughter, of whom I am fiercely protective.

I like to surround myself with out-of-the-box financiers and attorneys who are creative and fearless (just like James Bond and Batman) and share my integrity and respect values. I believe in due diligence, data, and not taking oneself too seriously.

Printed in Great Britain
by Amazon